# Rocks and Geologic History of the Pacific Northwest

## Geology Lab Manual

Written and compiled by
Autumn Christensen, Reanna Camp-Witmer,
and Shannon Othus-Gault

with additional contributions from Sheila Alfsen, Andrew Frank,
Bill Orr, and Mariah TIlman

Rocks and Geologic History of the Pacific Northwest: Geology Lab Manual
ISBN: 978-1-955499-42-2
© 2017, 2018, 2025 by Chemeketa Community College.

Chemeketa Press
Chemeketa Community College
4000 Lancaster Dr NE
Salem, Oregon 97305
collegepress@chemeketa.edu
chemeketapress.org

A full list of credits appears on page 75 and constitutes an extension of this copyright page.

References to website URLs were accurate at the time of writing. Neither the author nor Chemeketa Press is responsible for URLs that have changed or expired since the manuscript was prepared.

Printed in the United States of America.

**Land Acknowledgment**
Chemeketa Press is located on the land of the Kalapuya, who today are represented by the Confederated Tribes of the Grand Ronde and the Confederated Tribes of the Siletz Indians, whose relationship with this land continues to this day. We offer gratitude for the land itself, for those who have stewarded it for generations, and for the opportunity to study, learn, work, and be in community on this land. We acknowledge that our College's history, like many others, is fundamentally tied to the first colonial developments in the Willamette Valley in Oregon. Finally, we respectfully acknowledge and honor past, present, and future Indigenous students of Chemeketa Community College.

# Contents

# 1 What Geologists See

## Purpose

An introduction to aspects of Oregon's geologic history.

## Materials

- ❑ Google Earth
- ❑ Rock samples provided by your instructor

# Part 1. Reviewing Images and Samples

Using the images and samples provided, answer the following questions.

1. In Google Earth, explore the state of Oregon. Compare and contrast the Oregon Coast Range with the Cascade Mountains. How are they different? How are they similar?

2. Explain the differences between these two mountain ranges. Why would they be so different despite being relatively close together?

# Part 2. Describing Geological Images

**Figure 1.1. The Ellensburg Formation.**

3.  Look at Figure 1.1 and describe what you see. If you know what it is, don't use geologic terms. Describe it as a layperson.

4. In Figure 1.1, there are several layers as well as an obvious break/crack. In Figure 1.1, number the layers. Then list those layers, and the crack, in order from first to have occurred to last to have occurred. Provide a brief explanation for your order.

5. Provide an explanation for the crack/break in Figure 1.1. What would have happened to the rock layers to create such a feature?

6. Discuss the image with your fellow students. After sharing and discussing your ideas, create a new list from the first event to the last. Explain your answer. If this list is exactly the same as before, explain why you chose to stick with it over the ideas of your classmates (or provide new insight to support your original idea).

# Part 3. Describing Rock Samples

7. Look at the two rock samples provided and describe each one. Note any similarities and differences between them. Don't use geologic terms; describe it as if you have absolutely no knowledge of geology.

8. These two rocks are related. What do you think that relationship is? How do you think each rock formed? Use your observations from the previous question to support your answer.

9. Join up with a partner or group and share your observations and ideas. Come up with a collective idea on how these rocks have formed and how they are related. How can you support your answer with the observations you have made?

## 2 | Plate Tectonics

## Purpose
To identify and distinguish between the various plate boundaries and calculate plate velocities.

## Materials
- ❑ Calculator
- ❑ General plate boundary map
- ❑ Hawaiian Islands/Emperor Seamount map
- ❑ Seafloor Ages map
- ❑ Seismology map
- ❑ Topography map
- ❑ Volcanoes map

# Part 1. Activity at Plate Boundaries

Observe Figures 2.1 and 2.2 in this lab and then answer questions 1–4.

1.  Look at the following locations and circle what earthquake depths are found at that spot (more than one is possible).

    **Southern California**
    | Shallow | Intermediate | Deep | Very deep |

    **Japan**
    | Shallow | Intermediate | Deep | Very deep |

    **Oregon and Washington**
    | Shallow | Intermediate | Deep | Very deep |

    **The Himalayas**
    | Shallow | Intermediate | Deep | Very deep |

**Middle of the Atlantic Ocean**

    Shallow                Intermediate              Deep             Very deep

2. Look at the following locations and circle what volcano activity is found at that spot (circle one).

**Southern California**

    No Volcanoes                Some Volcanoes             Lots of Volcanoes

**Japan**

    No Volcanoes                Some Volcanoes             Lots of Volcanoes

**Oregon and Washington**

    No Volcanoes                Some Volcanoes             Lots of Volcanoes

**The Himalayas**

    No Volcanoes                Some Volcanoes             Lots of Volcanoes

**Middle of the Atlantic Ocean**

    No Volcanoes                Some Volcanoes             Lots of Volcanoes

3. Based on your previous answers, and the additional information from the other maps (age of the ocean floor and bathymetry/topography) what plate boundaries are located at the following locations? Explain your answers.

**Southern California**

    Convergent-Subduction        Convergent-Collision        Divergent        Transform

Explanation:

**Japan**

    Convergent-Subduction        Convergent-Collision        Divergent        Transform

Explanation:

**Oregon and Washington**

    Convergent-Subduction       Convergent-Collision      Divergent      Transform

Explanation:


**The Himalayas**

    Convergent-Subduction       Convergent-Collision      Divergent      Transform

Explanation:


**Middle of the Atlantic Ocean**

    Convergent-Subduction       Convergent-Collision      Divergent      Transform

Explanation:


4. Off the coast of Oregon we have a subduction zone, where the Juan de Fuca Oceanic Plate is going underneath the North American Continent. How is the subduction zone we see in the Pacific Northwest different from others around the Pacific? Suggest a reason for this difference.

# Part 2. Rates of Plate Movement

Observe the map of the Hawaiian Island and Emperor Seamount Chains (Figure 2.1) and answer the questions below.

**Figure 2.1. Hawaiian Island and Emperor Seamount Chain comparison.**

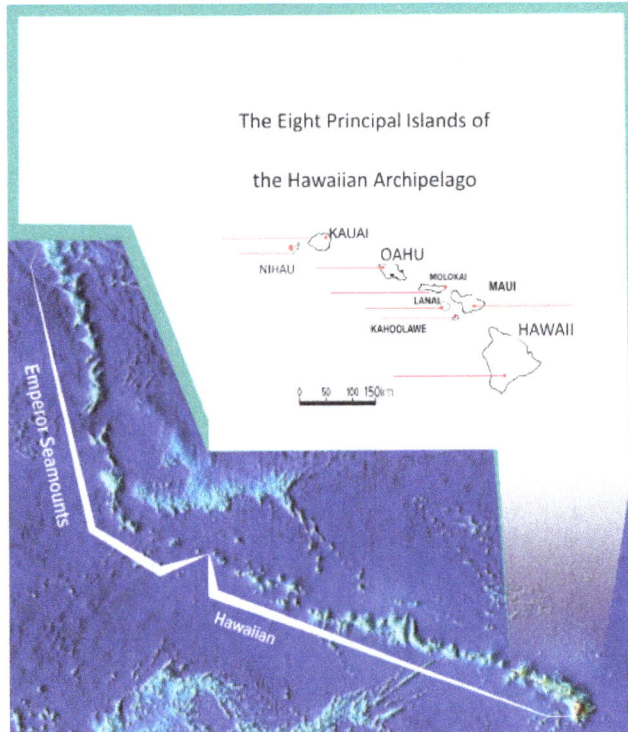

5. Looking at our maps of earthquake and volcanic activity, describe what activity we see in Hawaii.

6.  Is this activity due to its proximity to a plate boundary? If it is, which boundary? If it's not, what is causing the activity at this location?

7.  Looking at the map of Hawaii (Figure 2.1), why do the older islands decrease in size the further they are from the hotspot (you may want to look up some of the ages for the islands to get a general idea)?

8.  Which way is the plate moving? Explain your answer.

9.  Now view the orientation of the Emperor Seamount Chain. Assuming they were created from the same source, what does this tell us about the movement of the Pacific Plate over the past few 10s of millions of years?

10. What are the minimum and maximum ages of the island of Kauai? When you look them up, be aware of the numbers. If you see "Ma," that means "mega anum" or "millions of years."

    a.  Minimum: _____
    b.  Maximum: _____

11. Using the scale on the map (Figure 2.1), determine the distance from the hot spot to the center of Kauai.

    a.    Distance:_____ kilometers

    b.    (Convert): _____ centimeters

12. Using the distances and ages from the questions above, calculate the maximum and minimum rates (velocities/speeds) of Pacific plate movement in centimeters per year. Be careful with your numbers here; we want the final answer to be in centimeters per year.

    a.    Maximum velocity: _____ cm/yr

    b.    Minimum velocity: _____ cm/yr

# Part 3. Rates of Plate Movement Close to Home

Observe the map of the age of the ocean floor around the Juan de Fuca Ridge (Figure 2.2) and answer the following questions.

**Figure 2.2. Age of the ocean floor around the Juan de Fuca Ridge.**

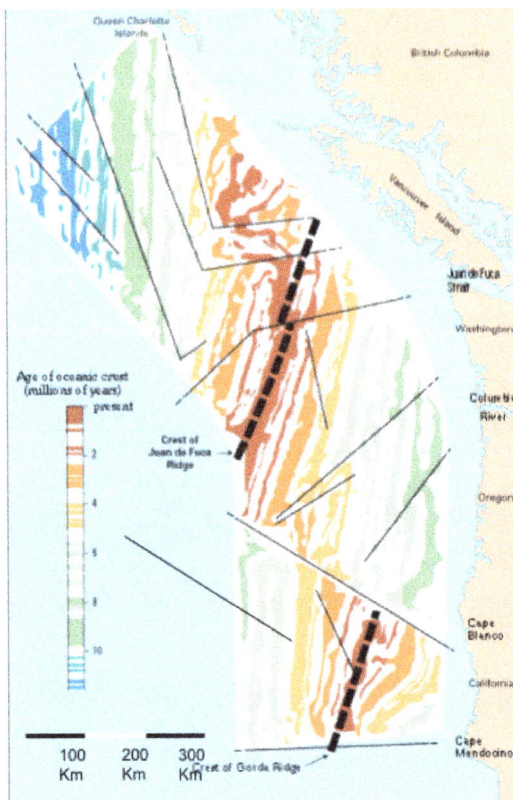

**13.** What type of rock dominates the ocean floor? Summarize the origin, texture, and composition.

**14.** What are the minimum and maximum ages of the ocean floor on this map?

    **a.** Minimum: _____

    **b.** Maximum: _____

**15.** Explain where you find the youngest rock on the ocean floor. Why is it located there?

**16.** Explain where you find the oldest rock on the ocean floor. Why is it located there?

**17.** Using the data provided from Figure 2.2, calculate the velocity of the Juan de Fuca Plate. Show all your work and explain your answer. Specifically, why are you using the numbers you're using?

# 3 Protoliths

## Purpose
Explore relationships between metamorphic rocks and their protoliths (parent rocks).

## Materials
- ❑ Glass plate
- ❑ Hand lens
- ❑ Hydrochloric acid
- ❑ Rock samples provided by your instructor

# Part 1. Common Metamorphic Minerals

While there are thousands of minerals in the world, most of the common metamorphic rocks contain a select few that we'll look at and learn to distinguish here.

*Instructions*
Using the provided mineral samples, fill in Table 3.1.

**Color**
While color can vary, some minerals have a more common appearance that helps distinguish them from others.

1. Using the provided samples, identify the color of each sample and fill in Table 3.1 accordingly.

**Mineral Hardness**
Hardness relates to a substance's ability to scratch another substance. In geology, we use the Mohs Hardness Scale. For metamorphic minerals, we're mainly interested in whether a mineral is hard, soft, or really soft.

2. Test the sample first with your fingernail. If the mineral can't scratch your fingernail (or you can make a scratch in the sample with your nail), then the sample is really soft.
3. If the sample scratches your fingernail, try to scratch the glass plate with it.

      a. If the sample cannot scratch the glass, it is a soft mineral.

      b. If the sample scratches the glass, it is a hard mineral.

4. Fill in Table 3.1 accordingly.

## Special Properties

Some minerals exhibit somewhat unique properties that easily distinguish them from other minerals. For metamorphic rocks, the most useful of these properties is whether the mineral reacts to acid.

5. If you have a soft mineral (scratches your fingernail but not glass), put a single drop of the provided acid on your sample. If it bubbles, it is the mineral calcite.
6. Wipe off your sample with a paper towel before filling out your table.
7. Once you have identified these properties, use Table 3.2 to identify your mineral (depending on the samples provided, you may have more than one example of some minerals).

**Table 3.1. Identification of Common Metamorphic Minerals**

| Sample Number | Color | Hardness | Acid Reaction | Mineral Name |
|---|---|---|---|---|
|  |  |  |  |  |
|  |  |  |  |  |
|  |  |  |  |  |
|  |  |  |  |  |
|  |  |  |  |  |
|  |  |  |  |  |
|  |  |  |  |  |
|  |  |  |  |  |
|  |  |  |  |  |
|  |  |  |  |  |

**Table 3.2. Common Metamorphic Minerals**

| Mineral Name | Color | Hardness | Other Properties |
|---|---|---|---|
| Calcite | Varies, often transparent or white | Soft | Reacts to acid, may have noticeable cleavage planes (flat geometric surfaces to the individual crystals) |
| Chlorite | Green | Soft to Very Soft | May appear platy/flakey, similar to micas |
| Feldspar | Varies, pink, white, or blue-grey most common | Hard | Blocky, rectangular. May have noticeable cleavage planes |
| Garnet | Dark Red, occasionally a light green | Hard | Very dark, resinous luster (shine). Often grows as dodecahedron crystals (twelve-sided ball) |
| Mica | Dark-translucent (Biotite) or Colorless/Yellow-white-Translucent (Muscovite) | Soft to Very Soft | Flakey/platy, easily peeled apart (please don't test that with your sample). Very reflective on the flat surface. |
| Serpentine Minerals | Most commonly green, sometimes with white or blue mixed in | Soft | Somewhat greasy feel. Often described as having the appearance of snakeskin (hence the name) |
| Talc | Varies, often white | Very Soft | Soapy/Greasy feel. Pearly luster (shine) |
| Quartz | Pick a color, any color | Hard | Glassy luster, may show conchoidal fracture (circular or shell-like pattern of breaking) |

# Part 2. Looking at Metamorphic Rocks

## Instructions

Using the provided metamorphic rock samples, fill in Table 3.3. For now, we're not worried about the name of the rock, just a general description

### Grain Size

1. Is the sample coarse-grained or fine-grained? Can you see the mineral crystals in it or not?

### Appearance and Texture

2. Describe the general appearance of the sample. Does the sample show any textures or patterns in the arrangement of the rock's minerals? If so, describe them.

### Identifiable Minerals

3. Is the sample made up of one mineral? Multiple minerals? If you can see the crystals, you can identify the minerals in the rock.

**Table 3.3. Metamorphic Rocks**

| Sample Number | Coarse or Fine Grained | Appearance and Texture | Identifiable Minerals |
|---|---|---|---|
| | | | |
| | | | |
| | | | |
| | | | |
| | | | |
| | | | |
| | | | |
| | | | |

# Part 3. Finding the Parent Rock

Your instructor will provide you with some common (but unidentified) sedimentary and igneous rocks. Looking at these samples and your metamorphic rocks from Part 2, answer the following questions.

3

1.  For each metamorphic rock provided, select a parent rock (protolith) that you think is the most likely origin for the sample. Provide an explanation for why and support your answer. "They look similar" is not enough of an answer (in some cases, rocks change considerably during metamorphism). Consider the minerals and textures of the parent rocks.

2.  Some of your parent rocks are composed almost entirely of one mineral. Others contain multiple minerals. How would this difference in mineral variety affect the resulting metamorphic products? Use examples from our samples.

3. For your parent rock/metamorphic rock sets composed solely of one mineral, what makes them different? If the original rock and the resulting changed rock are completely the same mineralogy, what changes occurred during metamorphism to make a supposedly new rock?

4. Based on your observations, what sorts of changes seem to occur during metamorphism besides potential changes to mineralogy?

5. Looking at your samples, is there any way you could tell which rock has undergone the most change from its original rock? Which one experienced the highest pressure and temperature changes? If so, how? If not, why not?

# 4 Metamorphic Rocks

## Purpose
To identify and distinguish between common metamorphic rocks.

## Materials
- ❑ Dilute hydrochloric acid
- ❑ Glass plate
- ❑ Hand lens
- ❑ Metamorphic lab samples

# Part 1. Describing the Properties of Metamorphic Rock

*Instructions*

For each sample, you will enter the answers to the following questions in Table 4.1.

1. **Texture**: Is the rock foliated or non-foliated?
2. **Grain Size**: Is the rock coarse or fine-grained?
   a. A rock is coarse-grained if you can see the crystals with the unaided eye or your hand lens.
3. **Mineralogy**: There is no specific answer for this, but it is a useful identification tool.
   a. Based on our limited mineral knowledge, what minerals do you think the rock is composed of?
4. **Features**: This is somewhat open; nothing is required here, but consider the following:
   a. *Does the sample react to acid?* For **non-foliated rocks**, a reaction suggests calcite is present.
   b. *Does the sample scratch glass?* For the **non-foliated rocks**, use your glass plate as a rough hardness test.
   c. Based on your responses to the previous questions you can discern whether the non-foliated rock is composed of calcite, serpentine, chlorite, talc, or quartz, which will help with identification.
      i. Calcite reacts to acid and does not scratch glass.

       ii.  Serpentine will not scratch glass.

      iii.  Quartz will scratch glass.

      iv.  Talc can be scratched by your fingernail.

      v.  Chlorite will not scratch glass and may or may not be scratched by your fingernail.

  d.  For foliated rocks, it is not important whether the sample reacts to acid or scratches glass, as neither will help with identification. It can, however, be helpful to identify minerals in the coarse-grained samples.

      i.  The most common metamorphic minerals in foliated rocks are quartz (glassy blobs), mica (platy, glittery), feldspar (blocky white or pink), and garnet (red balls).

5. **Rock Name**: Use the provided Table 4.2 to name the rock. There may be multiples, or some samples may not be present in our kits.

6. **Protolith**: For your chosen rock, list the protolith (or protoliths) that the rock would have originally formed from.

**Table 4.1. Metamorphic Rock**

| ID | Texture | Grain Size | Mineralogy | Features | Rock Name | Protolith |
|----|---------|-----------|-----------|----------|-----------|-----------|
| 1  |         |           |           |          |           |           |
| 2  |         |           |           |          |           |           |
| 3  |         |           |           |          |           |           |
| 4  |         |           |           |          |           |           |
| 5  |         |           |           |          |           |           |
| 6  |         |           |           |          |           |           |
| 7  |         |           |           |          |           |           |
| 8  |         |           |           |          |           |           |
| 9  |         |           |           |          |           |           |
| 10 |         |           |           |          |           |           |
| 11 |         |           |           |          |           |           |
| 12 |         |           |           |          |           |           |
| 13 |         |           |           |          |           |           |
| 14 |         |           |           |          |           |           |
| 15 |         |           |           |          |           |           |
| 16 |         |           |           |          |           |           |
| 17 |         |           |           |          |           |           |
| 18 |         |           |           |          |           |           |
| 19 |         |           |           |          |           |           |
| 20 |         |           |           |          |           |           |

4

**Table 4.2. Metamorphic Rock Identification**

| Texture | | | Mineralogy | Rock Name | Parent Rock |
|---|---|---|---|---|---|
| **Foliated** | Fine grained (not visible) | Appearance of dense mud | Mica, chlorite, quartz | Slate | Mudstone or shale |
| | | Metallic luster | Mica , chlorite quartz, garnet | Phyllite | Mudstone, shale, slate |
| | Coarse grained (visible) | Crystalline Texture | Mica, chlorite, quartz, garnet | Schist | Mudstone, shale, slate, phyllite |
| | | Crystalline Texture | Chlorite, epidote, actinolite | Greenschist | Basalt |
| | | Crystalline Texture | Glaucophane | Blueschist | Basalt |
| | | Banded | Feldspar, mica, amphibole, quartz, garnet | Gneiss | Mudstone, shale, slate, phyllite, schist, granite, diorite |
| **Either Foliated or Non-Foliated** | Coarse grained | Dark green or black | Amphibole | Amphibolite | Basalt, gabbro |
| | | Red and Green | Pyroxene, Garnet | Eclogite | Basalt, gabbro |
| | Varying grain size | Greasy, smooth feel | Talc | Soapstone | Basalt, gabbro, peridotite, dunite, serpentinite |
| **Non-Foliated** | | Black, glassy | Organic material | Anthracite | Peat, lignite, bituminous coal |
| | | Microcrystalline, usually dark | Chlorite, mica | Hornfels | Various |
| | | Smooth texture, like snake skin. Usually green | Serpentine minerals | Serpentinite | Basalt, gabbro, peridotite, dunite |
| | | Crystalline, color, varies | Calcite | Marble | Limestone |
| | | Sugary/sandy appearance, color varies | Quartz | Quartzite | Sandstone, conglomerate, chert |
| | | Fine grained | Chlorite, actinolite, epidote | Greenstone | Basalt |

# 5 Geologic Time

## Purpose
To explore the geologic timescale and practice basic half-life problems.

## Materials
- ❑ A length of receipt paper at least 5 feet long
- ❑ Calculator
- ❑ Colored pencils

# Part 1. Timescale Instructions

You will be making a geologic time scale using materials provided by your instructor. You won't be covering the entire timescale, just a small portion.

Based on the total length of your timescale, determine the scale (i.e., how many millions of years per inch).

### Instructions
1. **Start**: The timescale will "start" at the beginning of the Mesozoic Era, about 250 million years ago (this is the "starting point" in the questions).
2. **Time divisions**: Label the beginning (the point closest to the starting point) of all the eras, periods, and epochs in the last 250 million years, including:
   a. Mesozoic and Cenozoic eras
   b. Triassic, Jurassic, Cretaceous, Tertiary, and Quaternary periods
   c. Paleocene, Eocene, Oligocene, Miocene, Pliocene, Pleistocene, and Holocene epochs
3. Include with the label the absolute ages for these boundaries as given in your textbook.
4. On your timescale label key points in Oregon history.
   a. 175 Ma Josephine Ophiolite
   b. 118 Ma Baker and Wallowa terranes
   c. 60 Ma Siletzia Terrane
   d. 54 Ma Clarno formation
   e. 35 Ma Western Cascade

f.   17 Ma Columbia River Basalts

g.   6 Ma High Cascades

h.   12,000 yrs Missoula Floods

i.   7700 yrs Mt. Mazama

1.   Based on the total length of your timescale, what is your scale in the following units?

How many millions of years per inch? _____

How many millions of years per foot? _____

How many millions of years per centimeter? _____

2.   How far from your starting point did you put the Cenozoic? *Use your scale—the answer is a distance, not how many years.*

3.   How far from your starting point are the High Cascades (according to your scale)?

4.   How far from your starting point is the Quaternary Period (according to your scale)?

5.   How far from your starting point is the Pleistocene Epoch (according to your scale)?

6.   Your timescale is only the last 250 million years, but the Earth is 4.6 billion years old. Why do you think we haven't included the earlier eras and eons? *(Besides the fact that it would be really long and harder to make...)*

7. If we made a time scale using the circumference of the Earth (40,000 km), how far from our starting point (in kilometers) would you have to walk before you come to the modern Holocene epoch at ~10,000 years? Assume one time around is 4.6 billion years, and you start at the beginning of Earth. Show your work. *Note: We're not using your original scale; for this question, we're making a new one.*

# Part 2. Practice with Half Life

Answer the following questions and show all your work.

8. Over a period of 66 hours, the quantity of a radioactive isotope is reduced from 24.0 grams to 3.0 grams.
   a. How many half-lives have gone by? _____
   b. How long is the half-life of the isotope? _____

9. Strontium-90 has a half-life of 28.8 years.
   a. If you start with a 112-gram sample of strontium-90 and decay it for 115.2 years, how many half-lives will go by? _____
   b. What fraction or percent of the Strontium-90 will remain? _____
   c. How many grams of Strontium-90 will remain?_____

10. A sample contains 175 grams of parent isotope and 2,625 grams of daughter isotope. Assuming that no daughter product was present when decay first started, and the parent isotope has a half-life of 40years, how many years old is the sample?

# Part 3. Applying Our Relative Time Methods

List the events in Figure 5.1 in order from oldest (bottom) to youngest (top). Use the letter labels as seen in the figure.

- ❑ Sometimes we cannot tell exactly when an event happened. If you find such an incident, use arrows to indicate its earliest and latest possible occurrences.
- ❑ For any unconformities, include with the label what kind of unconformity it is.

**Figure 5.1. Geologic cross section.**

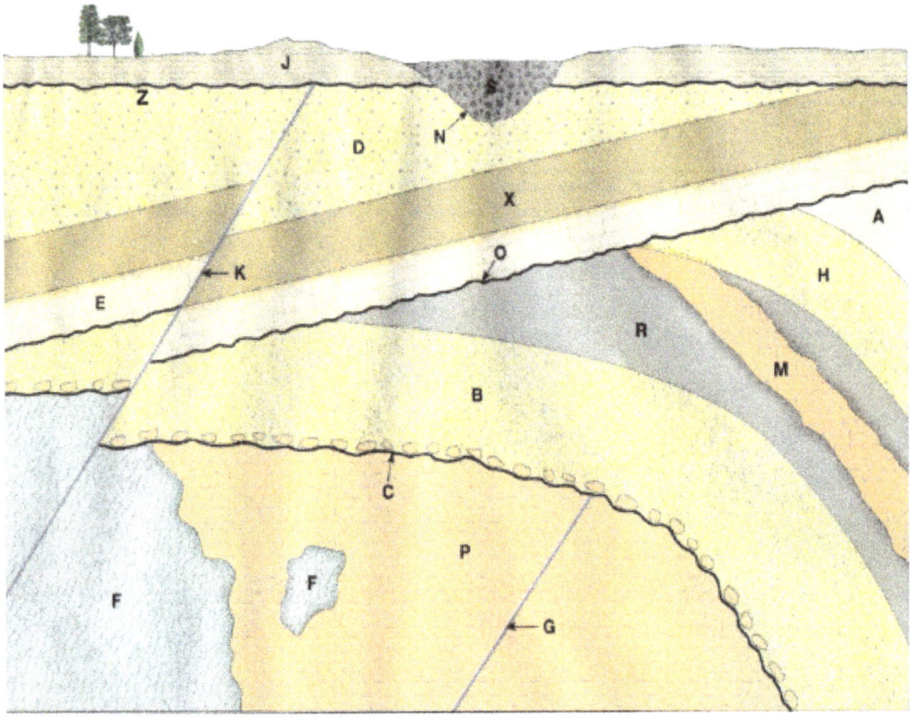

**Youngest**

_____
_____
_____
_____
_____
_____
_____
_____
_____
_____
_____
_____
_____
_____
_____
_____
_____
_____
_____

**Oldest**

11. Provide a summary explanation of why you went with the order you did. What principles are we following? Why did you choose to put one event before another? You don't need to be so detailed as to explain every step; if a group of layers is following the same rules and order, you combine them in your explanation. For example (not based on our diagram): "layers A-E are in order based on the principle of superposition, and Y comes after them as it cuts through all of those layers". Use another page of paper if you need the space.

**12.** For any unconformities, explain your choice. Why did you label it as that type of unconformity?

5

# 6 | Fossils

## Purpose
To identify fossil shells from Oregon.

## Materials
- ❑ Fossils provided by your instructor; (Alternate: fossil images provided by your instructor)
- ❑ Hand lens
- ❑ Identification booklets

# Part 1. Identifying Some Oregon Fossils

In this procedure, you'll identify and describe invertebrate fossils.

### Instructions
1. Look through the various samples or images provided.
2. Select six fossils to identify.
3. Using the provided materials, go through the various example images and identify the six fossils you have selected.
4. Fill out the provided pages (pp. 30–32) for each sample (1–6), including:
   a. The species name, generally given in italics.
   b. An explanation as to why you selected that particular species name. What specific features of your sample have led to this identification?
   c. A sketch of your sample.

### Notes
- ❑ Artistic ability isn't the goal here; detail is. Someone should be able to look at our collection of fossils and figure out which sample you're looking at based on your sketch.
- ❑ You are sketching the selected sample, imperfections and all (so not examples from our ID booklets).
- ❑ Your sample should be to scale. If you have shrunk or enlarged the image, you should indicate what magnification (e.g., ×2 would indicate you've drawn the sample twice as large as it actually is).
- ❑ If you're concerned about your drawings not being clear, feel free to use labels to indicate what things are, or include multiple views (e.g., top and side).

| Sample 1 | Image Number (if using): |
|---|---|
| Fossil Name | |
| Explanation | |
| Sketch | |

| Sample 2 | Image Number (if using): |
|---|---|
| Fossil Name | |
| Explanation | |
| Sketch | |

6

| Sample 3 | Image Number (if using): | Sample 4 | Image Number (if using): |
|---|---|---|---|
| Fossil Name | | Fossil Name | |
| Explanation | | Explanation | |
| Sketch | | Sketch | |

6

| Sample 5 | Image Number (if using): |
|---|---|
| Fossil Name | |
| Explanation | |
| Sketch | |

| Sample 6 | Image Number (if using): |
|---|---|
| Fossil Name | |
| Explanation | |
| Sketch | |

6

# Part 2. Not Identifying Some Oregon Fossils

1. Select four more samples from the available materials that you would consider difficult (or impossible) to identify.
2. Fill out the provided pages (pp. 34–35) for samples 7–10, including:
   a. An explanation of why these would be difficult or impossible to identify. What about the sample may prevent an accurate identification based on our provided materials? Consider what methods you used to make your identifications in Part 1.
   b. A sketch of each sample using the same criteria as Part 1.

6

| Sample 7 | Image Number (if using): |
|---|---|
| Fossil Name | |
| Explanation | |
| Sketch | |

| Sample 8 | Image Number (if using): |
|---|---|
| Fossil Name | |
| Explanation | |
| Sketch | |

6

| Sample 9 | Image Number (if using): | | Sample 10 | Image Number (if using): |
|---|---|---|---|---|
| Fossil Name | | | Fossil Name | |
| Explanation | | | Explanation | |
| Sketch | | | Sketch | |

6

# 7 | Structures

## Purpose

In this exercise, you will explore geologic maps, which are the primary means of communicating geologic information. You will investigate the primary types of crustal deformation (folds and faults), the types of stress that create them, and how they are represented on geologic maps and cross-sections.

## Materials

- ❑ Colored pencils
- ❑ Protractor
- ❑ Ruler

7

# Part 1. Strike and Dip on a Geologic Map

Examine the geologic map and its key (Figure 7.1).

**Figure 7.1. Geologic map for Part 1.**

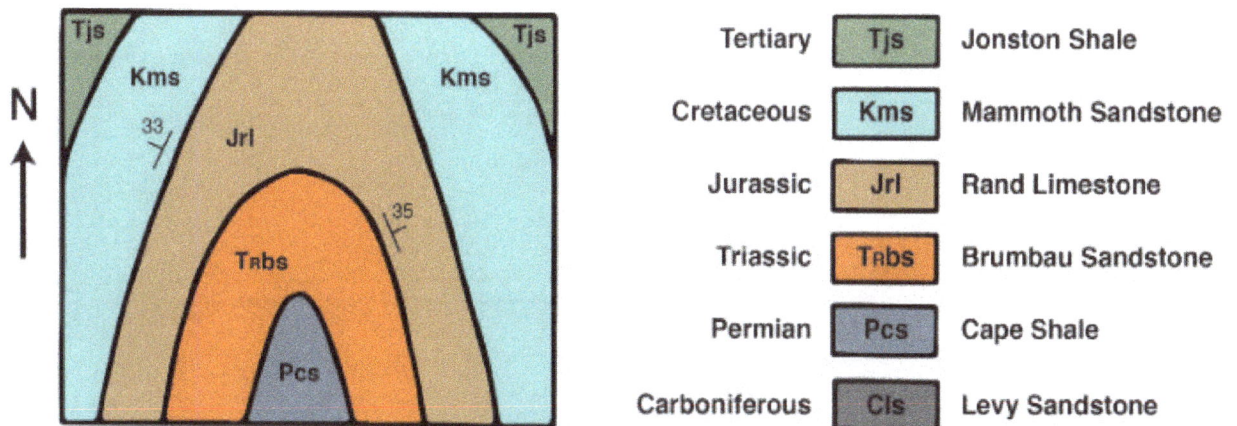

1.  What type of rock does the symbol Jrl represent?

2.  What does the "J" in Jrl stand for?

3.  What does the "rl" in Jrl stand for?

4.  Locate a contact on this map and name the formations on either side.

5.  Based on the strike and dip symbols on the map (Figure 7.1), what type of geologic structure is shown in this map?

Use Figure 7.2 to answer the following questions. For each lettered strike and dip symbol on the map, write out the approximate numeric strike and dip.

Figure 7.2. Map of strike and drip symbols for Part 1.

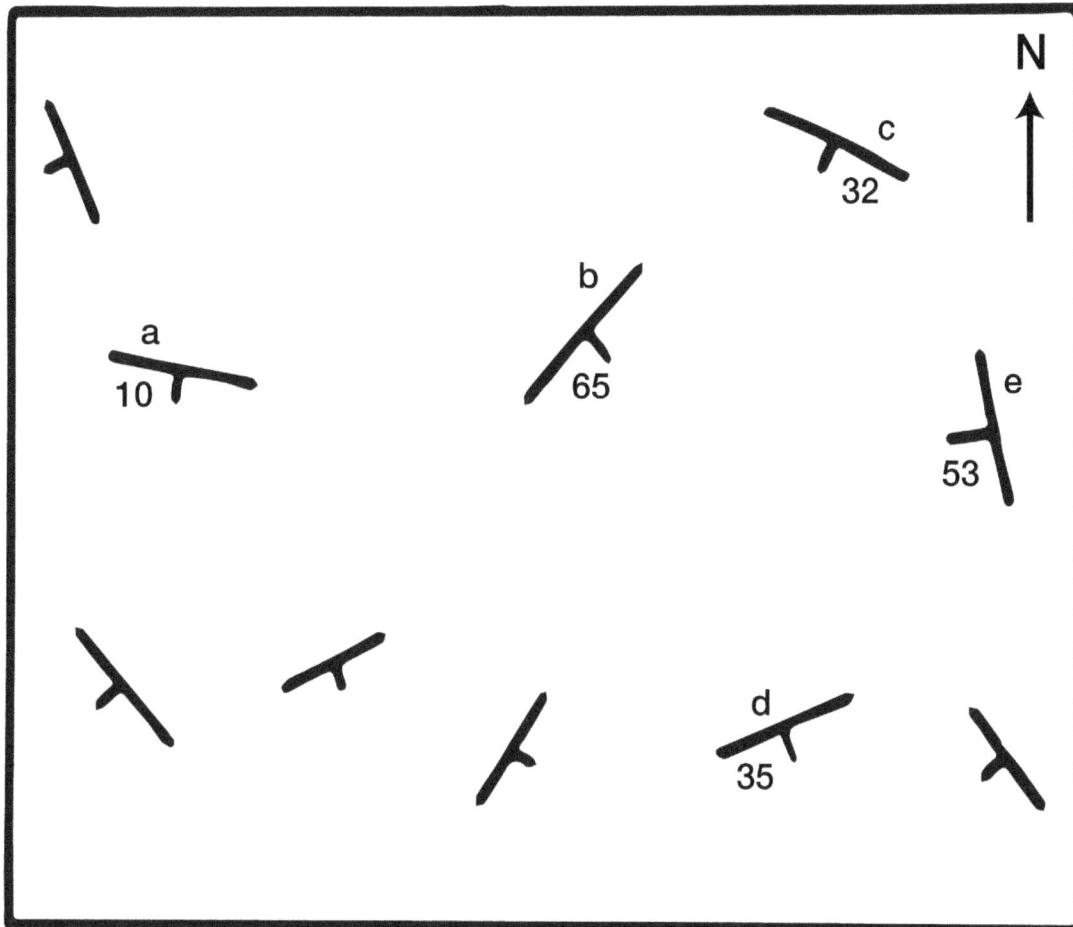

# Part 2. Block Diagrams

Using a pencil, complete all blank sides and the tops of block diagrams a-f in Figure 7.3. Add appropriate strike and dip symbols (if not already present) along the top, indicating the type of structure.

Letter labels indicate age based on a geologic timescale.

- ❑ S-Silurian
- ❑ D-Devonian

- ❑ M-Mississippian
- ❑ P-Permian

**Figure 7.3. Sketch strike and dip symbols on box diagrams.**

A.

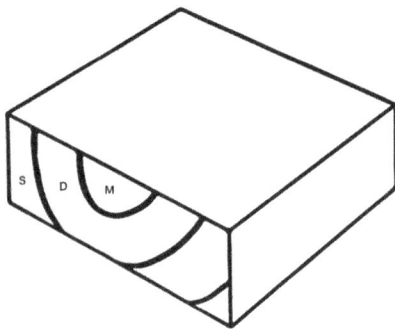

B.

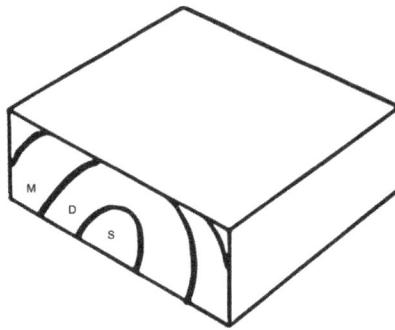

C.

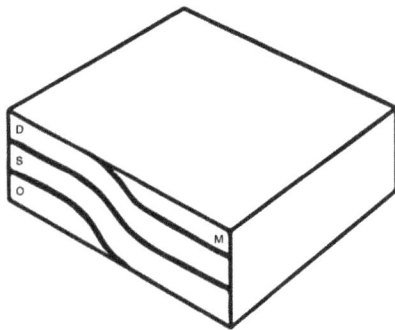

D.

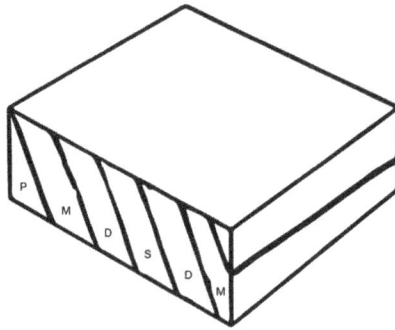

E.

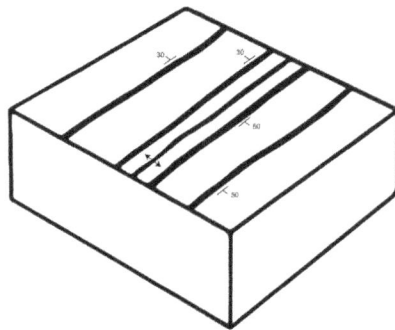

F.

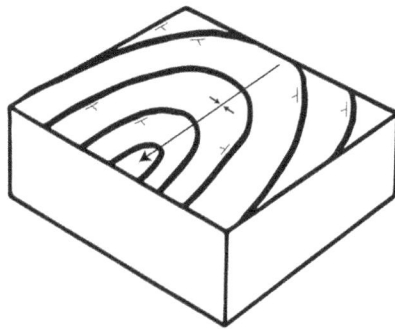

# Part 3. Cross Sections

Complete the cross sections A and B in Figure 7.4 using maps A and B. Answer the follow-up questions. *Note: The lower the number, the older the layer (number 1 being the oldest).*

**Figure 7.4. Reviewing and labeling cross sections.**

**Map A.**

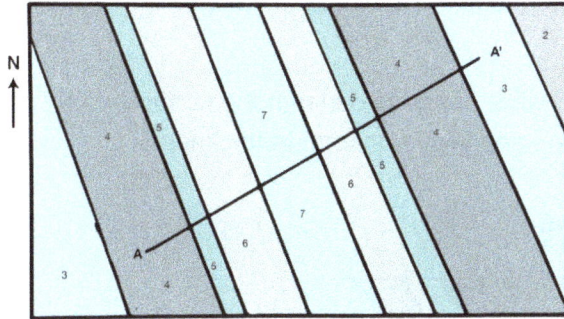

**Map B.**

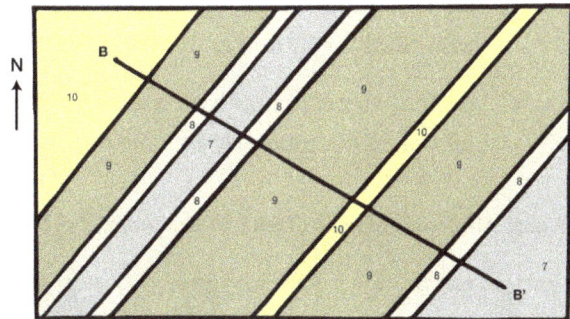

**Cross Section A.**

**Cross Section B.**

6. What type of fold is present in Cross Section A? How do you know?

7. What type of fold is present in Cross Section B? How do you know?

# Part 4. A More Complicated Cross Section

Complete the cross section in Figure 7.5. Answer the follow-up questions.

> **Note**
> - The irregular line provided on the graph is the surface.
> - The straight line is only for reference to measure angles. It is not the surface, and your cross section should just draw right over it.
> - You may need to make an assumption about what happens beneath the surface—follow the pattern already established and try to keep layers uniform in thickness.

**Figure 7.5. Cross sections of Silurian and Ordovician formations.**

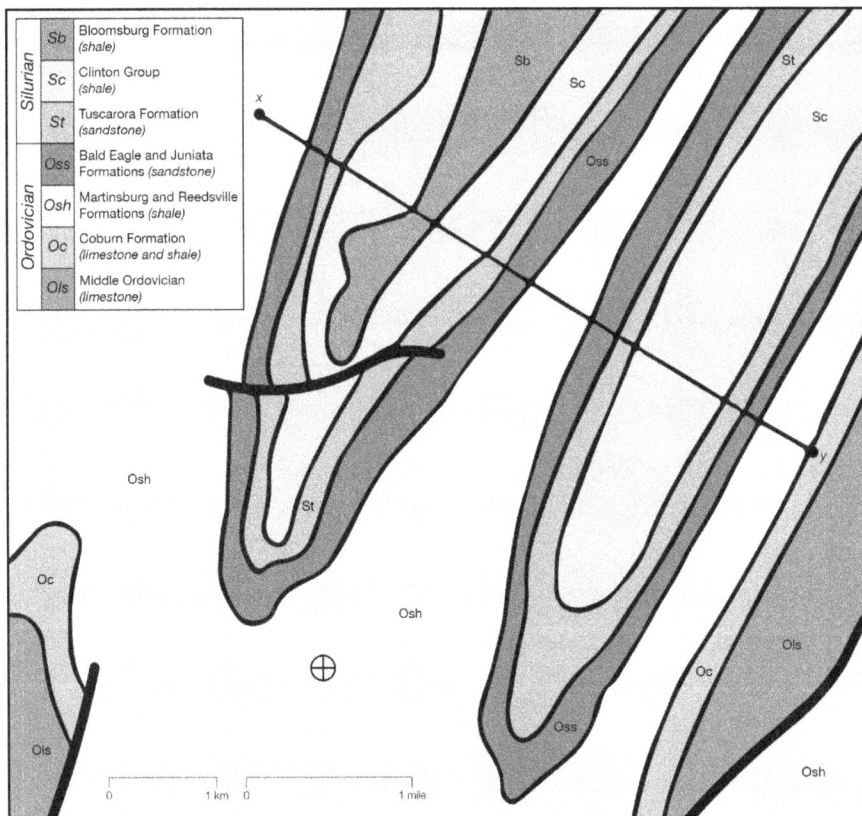

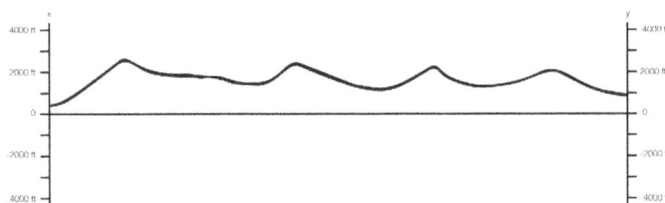

8.  Based on the outcrop pattern formed by the Silurian rocks in the area, what kind of structure do these rock types form?

9.  What is the general compass direction of strike for unit Oss?

10. What is the compass direction of dip for unit Oss where 50° dip is indicated?

11. Look at the fault in the approximate center of the map area. Did this fault form before or after the Silurian? Explain how you know.

# Part 5. Faults

For each fault image provided in Figure 7.6A–E, identify the type of fault and stress responsible for its creation. On each fault image in Figure 7.6A–E, draw arrows indicating relative motion along the fault, and label any hanging walls and footwalls.

Figure 7.6. Identifying faults in box diagram cross sections.

A.

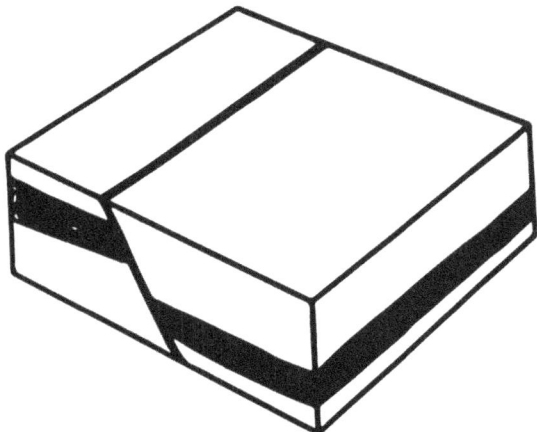

Fault and Stress: _____

B.

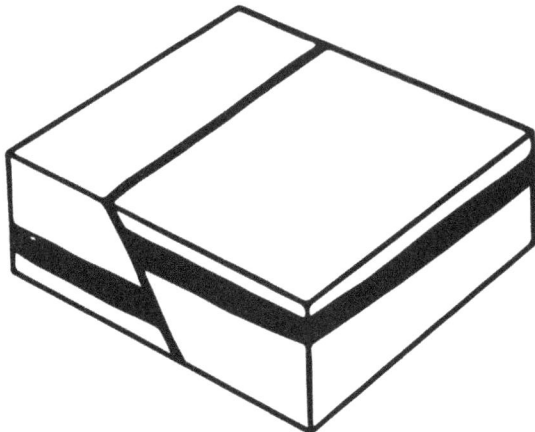

Fault and Stress: _____

C.

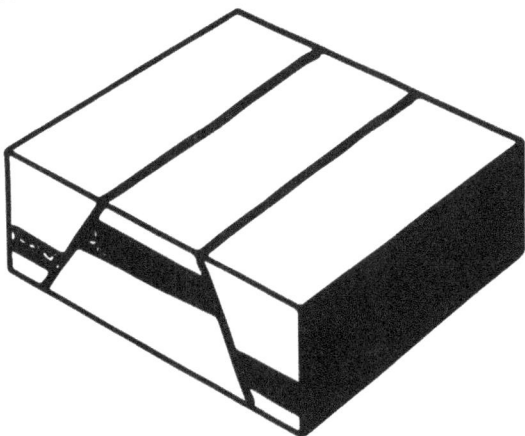

Fault and Stress: _____

D.

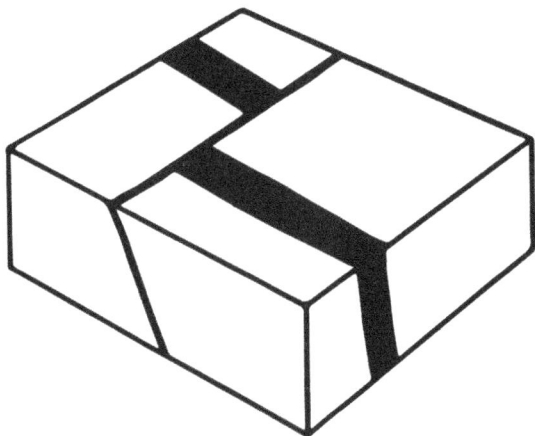

Fault and Stress: _____

E.

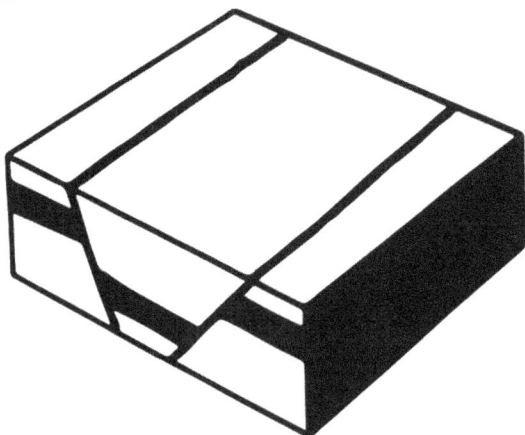

Fault and Stress: _____

# 8 | Geologic Map of Oregon

## Purpose
To explore a geologic map of Oregon.

## Materials
❑ Geologic map of Oregon

### Instructions
Answer the following questions. If asked for a rock unit or rock type, include the label and name of the formation as provided by the map legend. You may need to reference your textbook for some information on various types of rock.

# Part 1. Structures

8

1. Draw the symbol used for normal faults on the map (just named "faults" in our legend). Looking around the map, where do we see them most commonly? What does this indicate about the stresses experienced by this area?

2. Draw the symbol for thrust faults on the map. Where do we most commonly see them? What does this tell us about the stresses experienced in this area?

3. Where do you find a lot of folds on the map? *Note: You may need to zoom in a bit to see this.* Do they occur more commonly with one of our fault types? If yes, which one?

# Part 2. Igneous Rocks

4. What rock units make up Cascade Volcanoes like Hood and Jefferson? Do some investigation into that type of volcanic rock. What does it tell us about the type of eruptions common to these volcanoes?

8

5. Which is older, the Western or the High Cascades? Based on the information provided on the map, how do you know this? *Note: A map of Oregon provinces is provided on the key of the geologic map.*

6. Are the Western Cascades composed of the same types of rocks as the High Cascades? Provide examples.

7. An isopach is a line on a geologic map that connects areas of equal thickness of a material such as ash from a volcanic eruption. The map includes isopaches from Mt. Mazama (Crater Lake) and Newberry Volcano. The lines can be difficult to see, but their appearance and the meaning of the numbers on them are included in the key (legend) of the map. Based on the key and lines on the map, what was the predominant wind direction when the eruptions took place?

8. What is the most abundant volcanic rock unit (a rock that erupted to the surface) found on the map? Where is it exposed?

9. What is the most abundant plutonic rock unit (plutonic rocks cooled as magma under-ground) found on the map? Where is it?

10. Which is older, the volcanic or plutonic rocks of Oregon? Based on the information provided on the map, how do you know this?

# Part 3. Metamorphic Rocks

11. Where are the majority of metamorphic rocks in Oregon?

12. How old are the majority of metamorphic rocks (what period or era are they from)?

13. Look through the types of metamorphic rocks present, what types of parent rocks (protoliths) created these? What do these suggest about the original surface environment for the original rocks? Provide some examples of units from the map.

8

14. Based on the types of metamorphic rocks you've found, what type of metamorphism created the metamorphic rocks of Oregon? What supports this?

15. Are there any metamorphosed rocks in the Coast Range? *Note: The Coast Range is not the entire coast; look at the province map in the legend.* Suggest a reason why or why not.

# Part 4. Sedimentary Rocks

**16.** Where are the oldest sedimentary rocks in Oregon? What kind of rocks are they? *Note: we don't want "metasediments" here. That implies they've been altered.*

**17.** Which unit symbol represents the most recent deposits (which time period or epoch)?

**18.** What is Qal? Where is it always found? You may need to look closely at the map here.

**19.** What is Qd? Where is it found? There are two regions of the state where it is common.

**20.** What is Qs? In the Willamette Valley, this unit represents some recent geologic history from what is known as the Missoula Floods. Investigate this event (technically, these events) and explain Qs's formation.

21. What is Tt? Where is it found? These were formed by what's called turbidity currents. What environment did the sediments initially form in?

22. If Tt now makes up several mountain peaks, what does that mean about the tectonic history of the area?

8

23. Do units in the Blue Mountains and Klamath Mountains suggest the two share a common history (ignoring Quaternary deposits)? Provide examples of units that support your answer. *Note: Although they may not be identical units, examine the rock types and ages.*

# 9 Accreted Terranes of the Pacific Northwest

## Purpose

In this lab you will be creating a map based on the major accreted terranes of the western coast of North America, mainly in Oregon.

## Materials

- ❑ Colored pencils
- ❑ Glue stick
- ❑ Scissors

The thing about accreted terranes is that, geologically, they are a mess. They have everything you could possibly imagine in them: sedimentary rocks, igneous rocks, metamorphic rocks, continental rocks, oceanic rocks, fossils, etc. In fact, they are so messy that the geology is often called a mélange (the French word for "mixture"). Today you are going to create your own map of these Pacific Northwest accreted terranes. You will also determine the type of accreted terrain based on rock assemblages.

# Part 1. Assembling Your Map

## Instructions

1. You will need to assemble your map by cutting out your accreted terranes (Figure 9.3, p. 57) and gluing them to the map (Figure 9.2, p. 56) based on the borders shown, the cities on the terranes, and the main map.
2. After your terranes are glued onto the main map, you will color them based on their age and names.
   a. IMT—Intermontane Superterrane: Middle to late Jurassic Accretion—BLUE
   b. OB—Omineca Belt: Middle to late Jurassic Accretion—RED
   c. KT—Klamath Terrane: Late Jurassic to Early Cretaceous—GREEN
   d. CPC—Coast Plutonic Complex: Middle Cretaceous—YELLOW
   e. EOT—Eastern Oregon Terranes: Middle Cretaceous—PURPLE
   f. INT—Insular Superterrane: Middle to late Cretaceous—LIGHT BLUE
   g. OT—Olympic Terrane: Tertiary—ORANGE
   h. CT—Coastal Terrane: Tertiary to present—BROWN

1. Using the geologic timeline, what is the approximate time of accretion for each of the terranes and volcanic arcs? Where appropriate, use a range of years (ex, 90–40 million).

   a. Intermontane Superterrane: _____

   b. OB—Omineca Belt:_____

   c. KT—Klamath Terrane:_____

   d. CPC—Coast Plutonic Complex:_____

   e. EOT—Eastern Oregon Terranes: _____

   f. INT—Insular Superterrane: _____

   g. OT—Olympic Terrane: _____

   h. CT—Coastal Terrane: _____

# Part 2. Using Rock Assemblages to Identify Terranes

Use the rock assemblage described in questions 2–9 to identify what type of terrane(s) compose the terranes you mapped. A definition is given below for each terrane type, labeled **A**, **B**, **C**, or **D**. After each described rock assemblage, there is a blank space to list the type of terrane(s) related to it.

## Terrane Definitions

A. **Ophiolite:** A section of oceanic crust that can contain (from bottom to top): ultramafic rocks, gabbro, sheeted dikes, pillow basalts, siliceous ooze

B. **Volcanic Island Arc:** From subduction zone volcanism. Includes erupted volcanic rocks intruded by numerous diorite, granodiorite, and granite bodies

C. **Oceanic Plateaus:** Elevated sections of the oceanic crust. Some rocks are limestone formed from coral reefs. Also contains submarine fans similar to those currently being formed on the continental slope off the west coast (turbidites).

D. **Continental Volcanic Arc:** From the subduction of an oceanic plate under the North American Continent. Features include extensive folding, faulting, and metamorphism due to convergence, widespread intrusion by granites.

# Identify These Terranes

Use the letters associated with the terrane definition in your answers (A, B, C, D). *Note: Sometimes these rock assemblages occur together, so there can be more than one answer, with the exception of within a volcanic continental volcanic arc (D).*

2. **Intermontane Superterrane.** Erupted volcanic rocks intruded by granitic rocks, and packets of limestone and coral fossils. _____

3. **Omineca Belt:** A widespread area of mainly granite and gneiss._____

4. **Klamath Terrane:** Basalt and andesite intruded by granites, limestone pockets, coral fossils, sheeted dikes, and pillow lavas. _____

5. **Coast Plutonic Complex:** Massive amounts of molten granite and schist were injected into this area. _____

6. **Eastern Oregon Terranes:** Volcanic rocks like andesite, intruded by numerous diorite, granodiorite, and granite bodies. Also contains sequences of volcanic sediments with limestone and fossils of coral reefs, sandwiched between pillow basalts, gabbros, and sheeted dikes. _____

7. **Insular Superterrane:** Erupted volcanic rocks intruded by granitic rocks and packets of limestone and coral fossils._____

8. **Olympic Terrane:** A lot of basalt, pillow basalts, and submarine fans (turbidites).

_____

9. **Coastal Terrane:** Gabbro and a large amount of basalt. Also contains turbidites.

_____

# Part 3. Identifying Accreted Terranes of the Pacific Northwest

10. In some of the accreted terranes of the Pacific Northwest, there are fossils of fusulinids, a type of primitive, single-celled animal that floated in ocean water. They flourished in tropical oceans during the late Paleozoic era, during the Mississippian through Permian periods. The fusulinid fossils in the PNW are called Tethyan fusulinids because they are types of fusulinids that existed in the large sea known as the Tethys Sea on the east side of the supercontinent Pangaea (Figure 9.1). Based on this information and the map (Figure 9.1), how did the terranes of the Pacific Northwest reach their current locations?

**Figure 9.1. Landform during the Triassic Era.**

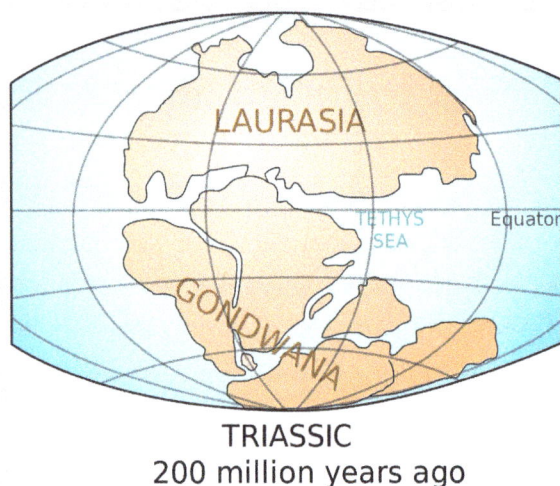

TRIASSIC
200 million years ago

11. There is a line that starts in northeastern Washington and then travels approximately along the border of Washington/Oregon, and Idaho. This line separates two areas with different Strontium-87 values. Strontium-87 is a stable isotope, meaning it is no longer radioactive. Its parent material is Rubidium 87. Rocks east of this line have a much higher concentration of Strontium 87, while rocks west of this line have a comparatively low concentration of Strontium 87. Explain why this occurs.

12. If you are walking on an ophiolite complex, it is a sure sign that you are walking on accreted terrane. How do you know you are walking on an ophiolite?

9

13. It is rare to see an entire ophiolite sequence in an accreted terrane. Give two reasons for why this is and explain your answer.

Figure 9.2. Accreted terrane map.

Accreted Terrane Map

Hand This In!

Port Angeles, WA

Vancouver B.C.

Las Vegas B.C.

Newport, OR

Victoria B.C.

Sand Point, ID

Medford, OR

Baker City, OR

**Figure 9.3.** Accreted terranes: cut and color these.

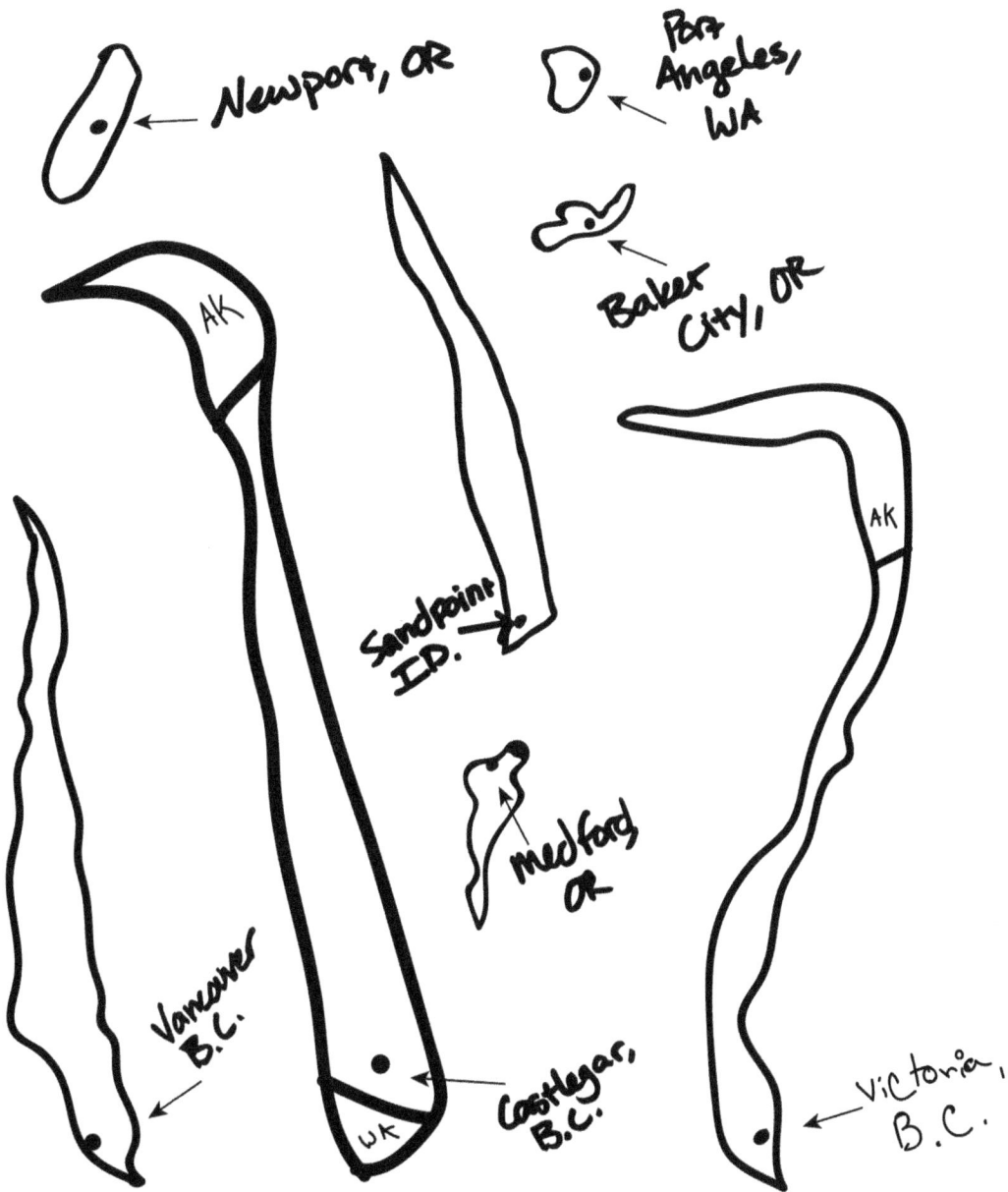

Accreted Terranes: Cut & Color These.

- Newport, OR
- Port Angeles, WA
- Baker City, OR
- AK
- Sandpoint ID.
- AK
- Medford OR
- Vancouver B.C.
- Castlegar, B.C.
- WA
- Victoria, B.C.

# 10  Field Trip: Downtown Salem Walking Tour

## Purpose

The purpose of this trip is to see a variety of rocks used as commercial building stone in Salem. Building stones are selected mainly for their beauty, and are not normally integral to the building structures. As you'll learn, many of these rocks were quarried from faraway places.

## Part 1. Begin Your Tour

Answer the following questions for each stop. For rock identification, use Tables 10.1–10.3. Begin at 220 Liberty Street NE, Salem.

## 248 Liberty Street NE
*(Formerly Café Shine)*

1.  Find a surface where the red paint on the bricks here has been removed. Is this rock igneous, sedimentary, or metamorphic?


2.  What is the name of this rock?


## 255 Liberty Street NE
*(Across the street from where you started)*

3.  Is this rock igneous, sedimentary, or metamorphic?


4.  What is the texture of this rock (look at the size of the mineral grains)?


5.  What is the composition of this rock (low, mid, or high silica)?


6.  What is the rock name?

## 241 Liberty Street NE

*This rock is fake. But what is it imitating...?*

7.  Is this "rock" igneous, sedimentary, or metamorphic?

8.  What is the texture of this "rock"?

9.  What is the "rock" name?

## 225 Liberty Street NE

*(Jackson's Jeweler)*

10. Is this rock igneous, sedimentary, or metamorphic?

11. What is the texture of this rock?

12. What is the composition of this rock?

13. What is the rock name?

14. This rock has several dikes running through it. Choose one and compare its texture and composition to the main rock.

## 399 Court Street NE

*(Bentley's Coffee)*

**15.** This rock is foliated. What does this mean?

**16.** What is the red mineral in this rock?

**17.** What is this rock's name?

## 377 Court Street NE

*(India Palace Restaurant Flower Planter)*

**18.** Are the bricks on the planter igneous, sedimentary, or metamorphic?

**19.** What is the rock name?

**20.** What previous stop do these rocks resemble?

## 179 Commercial Street NE

**21.** Look down at the sidewalk. Notice the blocks of purple glass in the pavement. Describe the fracture pattern these blocks exhibit.

**22.** This pattern is known as conchoidal fracture. It's common to some rocks and minerals you are likely familiar with. Can you think of which ones?

## 129 Commercial Street NE

This is an old rock, at least half a billion years old.

23. Is it igneous, sedimentary, or metamorphic?

24. What is the rock name?

25. What is the mafic (dark, low silica) mineral in this rock?

26. What is a possible parent rock?

## 109 Commercial Street NE

*(Pioneer Trust Bank)*

This rock is a good example of what's called the Schiller effect, which is the flash of color you see across the crystals.

27. Which feldspar is present in this rock?

28. Move around so you catch different angles of light reflecting off the rock. What do you see?

## 379–383 State Street NE

*(Ma Valise Jewelry Store, back towards Liberty)*

29. Is this rock igneous, sedimentary, or metamorphic?

30. This rock is soft, but does not react to acid. What is the rock name?

31. This rock is ultramafic. What is the parent rock?

# 416 State Street NE

*(Key Bank)*

The bricks in this building are composed of two types of limestone: coral limestone and travertine (nonmarine limestone, which forms in caves).

**32.** How are the two limestone types similar? How are they different?

**33.** What evidence (visible in the bricks themselves) can you find that indicates these bricks are composed of calcite?

# 512–516 State Street

*Corner of State and High Streets (The Marion County Courthouse)*

This rock is made out of the same mineral as the previous stop, but it is not limestone.

**34.** What is the rock name?

**35.** Look closely (use a magnifying glass or hand lens if you have one). What light-colored mineral forms thin streaks in the rock?

**36.** There is a metallic mineral present here. What is it?

# 900 State Street

*In front of Collins Science Center at Willamette University, an erratic from Missoula Floods with plaque*

This boulder was deposited in Oregon 11,500–13,000 years ago (the plaque is lying to you) during the Missoula Floods.

**37.** Is it igneous, sedimentary, or metamorphic?

**38.** What is the rock name?

**39.** What is the mafic mineral in this rock?

# 1040 State Street

*Eaton Hall at Willamette University*

The foundation of this building is a tuffaceous sandstone.

**40.** How is the weathering of this rock different from the sandstone at Stop #1?

**41.** Why would the weathering of this rock be different if they are both sandstone?

# State Capitol Park, Circuit Rider Statue
*East side of Capitol Building*

42. Is this rock igneous, sedimentary, or metamorphic?

43. Does it contain quartz? You may see this better if the rock is wet; they look like grey blobs of glass.

44. What is the mafic (dark mineral) in the rock?

45. What is the rock name?

# State Capitol Park, Jason Lee and/or John Mclaughlin Statues
*East side of Capitol Building*

10

46. What texture do you see here that was not at the Circuit Rider statue?

# State Capitol Park, Breyman Fountain
*West side of park, next to Cottage Street NE*

The pavers around the fountain are made of limestone. Some of them are travertine, as we saw at Stop #11, Key Bank.

**47.** What feature distinguishes the travertine bricks from the other limestone?

**48.** Notice some of the limestone bricks have a zig-zag pattern running through the rock. Suggest an origin for this feature.

**Table 1.10. Igneous ID Table**

| Texture | High Silica<br>Lighter colors | Mid Silica<br>Mix of light and dark dolors | Low Silica<br>Mostly dark colors |
|---|---|---|---|
| **Fine Grained**<br>*no visible crystals* | Rhyolite | Andesite | Basalt |
| **Two-Grained**<br>*like a chocolate chip cookie* | | | |
| **Coarse Grained**<br>*visible crystals* | Granite | Diorite | Gabbro |

**Table 10.2. Sedimentary Rocks**

| Texture | Particle Size (detrital) or Composition (chemical) | Rock Name |
|---|---|---|
| **Detrital** *physically broken minerals and rock fragments* | Clay *no visible grains* | Shale |
| | Silt *grains not visible, but has a gritty feel* | Siltstone |
| | Sand *visible grains* | Sandstone |
| | Gravel *visible grains* | Conglomerate *rounded pieces* **or** Breccia *angular pieces* |
| **Chemical** *crystalline of shell material* | Calcite | Limestone |
| | Quartz | Chert |

**Table 10.3. Metamorphic Rocks**

| Texture | Visible Minerals (foliated) *or* Composition (non-foliated) | Rock Name |
|---|---|---|
| **Foliated** *aligned minerals* | No | Slate |
| | No *metallic luster* | Phyllite |
| | Yes | Schist |
| | Yes *banded/striped* | Gneiss |
| | Yes *swirls and stripes, formed from partial melting* | Migmatite |
| **Non-foliated** | Calcite | Marble |
| | Quartz | Quartzite |
| | Serpentine Minerals | Serpentinite |

10

**Table 4: Common Mineral Descriptions**

| Mineral Group | Mineral | Description | Common Rock Types |
|---|---|---|---|
| High Silica Minerals | Quartz | Often presents as gray, glassy blobs | • High Silica rocks like granite or rhyolite<br>• Detrital Sedimentary Rocks<br>• Chert |
| | Potassium Feldspar | Rectangular, often pink | • High Silica rocks like granite or rhyolite |
| | Muscovite | Flaky, almost like glitter, light colors or transparent | • High silica rocks like granite or rhyolite<br>• Foliated metamorphic rocks |
| | Biotite | Flaky, almost like glitter, brown to black | • Igneous rocks like granite or diorite (or their volcanic equivalents)<br>• Foliated metamorphic rocks |
| Low Silica Minerals | Amphibole | Longer than they are wide, often look like splinters of wood, black | • Igneous rocks like andesite or diorite<br>• Foliated rocks like gneiss |
| | Pyroxene | Square, often dark brown or green | • Igneous rocks like peridotite, basalt, or gabbro |
| | Olivine | Looks like green glass | • Igneous rocks like peridotite, basalt, or gabbro |
| Other Silicates | Garnet | Red balls (dodecahedrons) | • Foliated metamorphic rocks |
| | Plagioclase Feldspar | Rectangular, ranges from white to dark blue or grey | • All igneous rocks<br>• Metamorphic rocks like gneiss |
| | Serpentine Minerals | Green, snakeskin appearance | • Serpentinite |
| Non-Silicates | Calcite | White, yellow, or colorless; reacts to acid | • Limestone<br>• Marble |
| | Magnetite | Small, dark, reflective crystals, possibly metallic | • Trace amounts in many rock types |
| | Pyrite | Fools gold, yellow/brass/gold color, metallic | • Trace amounts in many rock types |
| | Galena | Silver colored, cubic crystals, metallic | • Trace amounts in many rock types |
| | Hematite | Metallic form looks like small silver pieces of glitter<br>Non-metallic form often looks like red clay | • Trace amounts in many rock types |

10

# 11 Field Trip: Oregon Coast

## Purpose

To explore areas of the Oregon Coast and see examples of Oregon's volcanic and tectonic history.

*Instructions*

At each stop, follow your instructor's directions and answer the following questions.

## Stop #1. Cascade Head

1. Describe and name the rocks/boulders along the boat launch. You should be able to find at least three different kinds. Focus on igneous, sedimentary, and metamorphic. If you can identify the rock further from there, please do so. Explain your answer. Why have you identified it as such?

2. Sketch the rock outcrop in the distance (indicated by your instructor). Based on the features you see (which should be visible in your sketch),. what rock unit is represented here? Which rock from question 1 is this most like?

**11**

3. Observe the river nearby. Is the river gradient relatively flat or steep? How does the river sit compared to the surrounding terrain? Is the area flat, hilly, canyon-like, steep, some combination thereof, or something else? Given what we know about this area of the coast, what is the relationship between the river and local tectonics?

# Stop #2. Proposal Rock, Neskowin

**4.** Look at the rock outcrop on Proposal Rock. Describe what you see. Which rock from stop #1 is this rock most like?

**5.** How is the basalt of Proposal Rock different from the basalt you've learned about previously in class? What's strange about it? Sketch the outcrop, highlighting those differences and oddities.

**6.** Suggest an origin for this rock. If it isn't just your regular run-of-the-mill basalt, what happened to it? Explain your reasoning.

**7.** After your instructor has provided an explanation for the material making up Proposal Rock, were you correct about its origins? What does the true origin of this rock indicate about volcanism in this location?

**11**

**8.** Depending on the tides, a series of stumps may be visible along the beach nearby. Suggest an origin for the sunken forest we're seeing. Explain your reasoning.

# Stop #3. Cape Kiwanda, Pacific City

9. Which rock from Cascade Head is the headland (the piece of coast sticking out into the ocean) most similar to? How is the headland rock visually distinct from the nearby dark rock besides color?

10. Which rock from Cascade Head is the darker rock most similar to?

11. One of these rocks makes up the majority of the headland; the other cuts through it and also makes up Haystack Rock in the distance. Explain their relationship. Which is older? Which is younger? How do we know?

# Stop #4. Mount Hebo

12. Describe the topography of the valley below. One concern in the Pacific Northwest is tsunamis after an earthquake. Given the terrain and our tectonic setting, why would tsunamis be a big concern in this area?

13. Explain the general shape of the major peaks in the Coast Range. Why are so many of them flat peaks? Provide a general order of events leading to their formation.

11

# Acknowledgments

## Image Acknowledgments

Unless otherwise stated, figures are created and copyrighted by the authors or Chemeketa Press. Copyright © 2025.

Figure 1.1. "Reading the Washington State Landscape," permission granted by author Ben McShane, 2025.

Figure 2.2. "Magnetic anomalies of west coast of North America," by W. Jacquelyne Kious and Robert I. Tilling (https://pubs.usgs.gov/gip/dynamic/Endnotes.html). Wikimedia Commons Creative License CC0.

Figure 9.3. "The continents Laurasia-Gondwana 200 million years ago," by Lennart Kudling. Wikimedia Commons Creative License 3.0.